THE WILD LIFE OF DINOSAURS

AND OTHER PREHISTORIC ANIMALS

BUSTER BOOKS

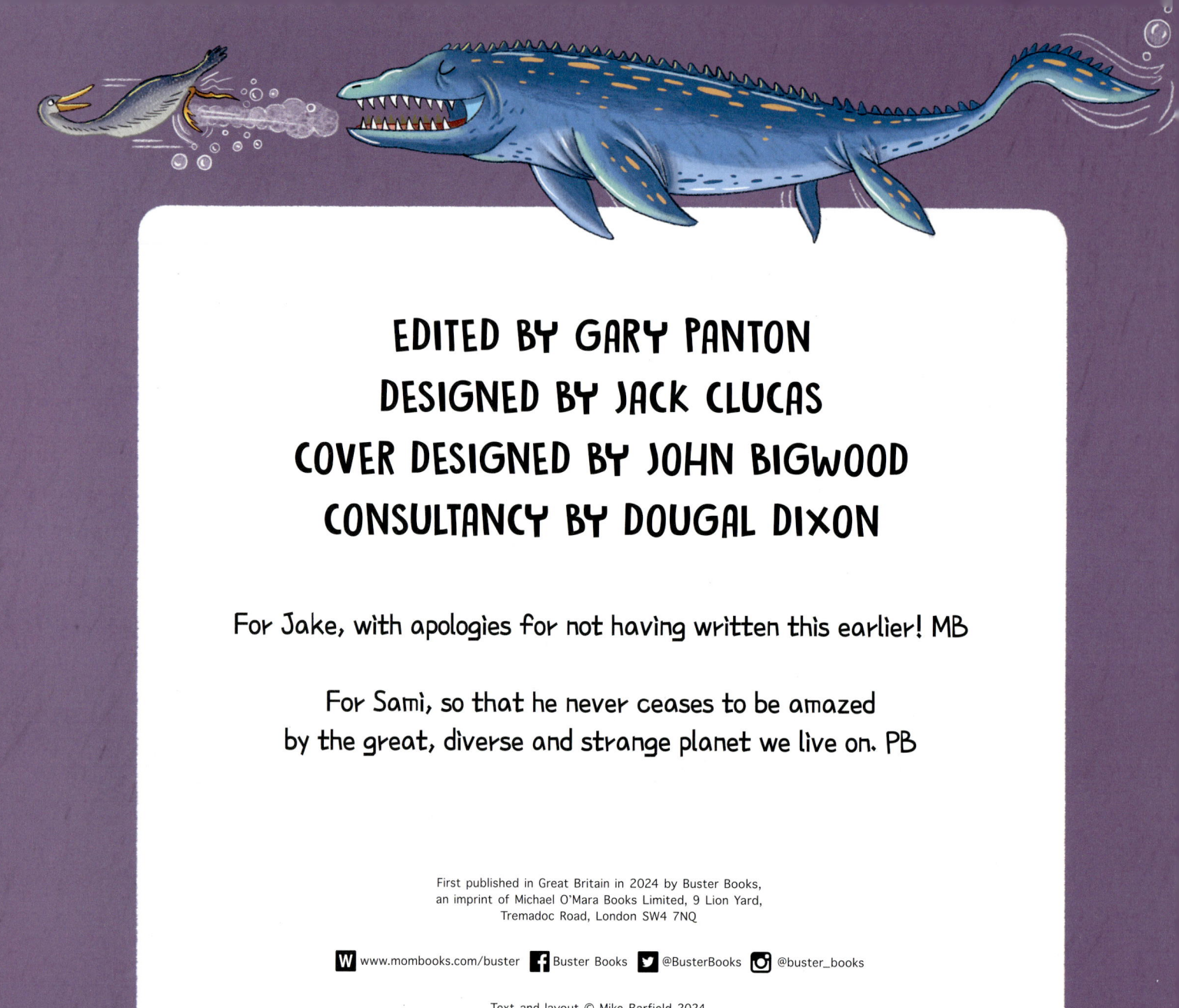

EDITED BY GARY PANTON
DESIGNED BY JACK CLUCAS
COVER DESIGNED BY JOHN BIGWOOD
CONSULTANCY BY DOUGAL DIXON

For Jake, with apologies for not having written this earlier! MB

For Sami, so that he never ceases to be amazed by the great, diverse and strange planet we live on. PB

First published in Great Britain in 2024 by Buster Books, an imprint of Michael O'Mara Books Limited, 9 Lion Yard, Tremadoc Road, London SW4 7NQ

W www.mombooks.com/buster   Buster Books   @BusterBooks   @buster_books

Text and layout © Mike Barfield 2024

Illustrations copyright © Buster Books 2024

All rights reserved. No part of this publication may be reproduced, stored in a retrieval system, or transmitted in any form or by any means, electronic, mechanical, photocopying, recording or otherwise, without prior permission of the publisher, nor be otherwise circulated in any form of binding or cover other than that in which it is published and without a similar condition including this condition being imposed on the subsequent purchaser.

A CIP catalogue record for this book is available from the British Library.

ISBN: 978-1-78055-932-2

1 3 5 7 9 10 8 6 4 2

This book was printed in April 2024 by Shenzhen Wing King Tong Paper Products Co. Ltd., Shenzhen, Guangdong, China.

THE WILD LIFE OF DINOSAURS

AND OTHER PREHISTORIC ANIMALS

WRITTEN BY MIKE BARFIELD

ILLUSTRATED BY PAULA BOSSIO

CONTENTS

INTRODUCTION	8
ENTER THE ANIMALS	9
Charnia	10
Living Legends: Jellyfish	11
Anomalocaris	12
Trilobite	14
Opabinia	16
Dead Cool: Cambrian Catwalk	17
Living Legends: Brittle Star	18
Dead Cool: Deep Dive	19
Eurypterid	20
Living Legends: Silverfish	22
Coelacanth	23
Dunkleosteus	24
Tiktaalik	26
Dead Cool: Go Fish!	27
Arthropleura	28
Living Legends: Mayfly	30
Hylonomus	31
Meganeura	32
Dead Cool: Fashionably Late	34
Diplocaulus	35
Dimetrodon	36
Dead Cool: Over and Out	38
DINOSAUR DAYS	39
Lystrosaurus	40
Living Legends: Tuatara	41
Coelophysis	42
Dead Cool: Bouncing Back	44
Leedsichthys	45
Allosaurus v *Stegosaurus*	46
Living Legends: Horseshoe Crab	48

Brachiosaurus	49
Diplodocus	50
Archaeopteryx	52
Rhamphorhynchus	53
Dead Cool: Jurassic Sparks	54
Plesiosaur	55
Amargasaurus	56
Spinosaurus	57
Pteranodon	58
Deinosuchus	59
Mosasaurus	60
Parasaurolophus	62
Velociraptor	63
Living Legends: A Sea Three	64
Carnotaurus	65
Ankylosaurus	66
Tyrannosaurus v *Triceratops*	68
Dead Cool: Audacious Cretaceous	70

AFTER THE ASTEROID — 71

Titanoboa	72
Coryphodon	73
Ambulocetus	74
Living Legends: European Eel	75
Archaeotherium	76
Megalodon	78
Phorusrhacos	80
Living Legends: Hoatzin	81
Dead Cool: The Fur Brigade	82
Australopithecus	83
Megatherium	84
Woolly Mammoth v Neanderthal	86
Dire Wolf	88
Dodo	89
Dead Cool: Life Lessons	90
Living Legends: Human	91

GLOSSARY — 92

TIMELINE — 94

INTRODUCTION

This isn't just a book — it's a time machine! Dip into the pages that follow and you'll travel back hundreds of millions of years to meet many of Earth's previous occupants.

'Prehistoric' means from a time before people were able to record or write things down. Before humans came along, there were all manner of weird and wonderful animals that we now know only from fossils — rocky clues that scientists use to picture the past.

Scientists are able to use fossils to guess at how prehistoric animals lived. However, many of these animals' characteristics and behaviours remain shrouded in mystery.

This book uses the best information available to bring some of those amazing animals back to life and tell their stories. Alongside those stories are 'Dead Cool' guides, catching up with additional early animals, and 'Living Legends' pages, highlighting the ancient species that are still with us today.

Yes, you'll meet dinosaurs, but you'll also meet killer shrimps, fearsome fish, walking whales and your own ape-like ancestors. So, sit down, strap in and set the time machine controls to 'prehistoric' — you're in for a wild ride!

ENTER THE ANIMALS: PALEOZOIC ERA

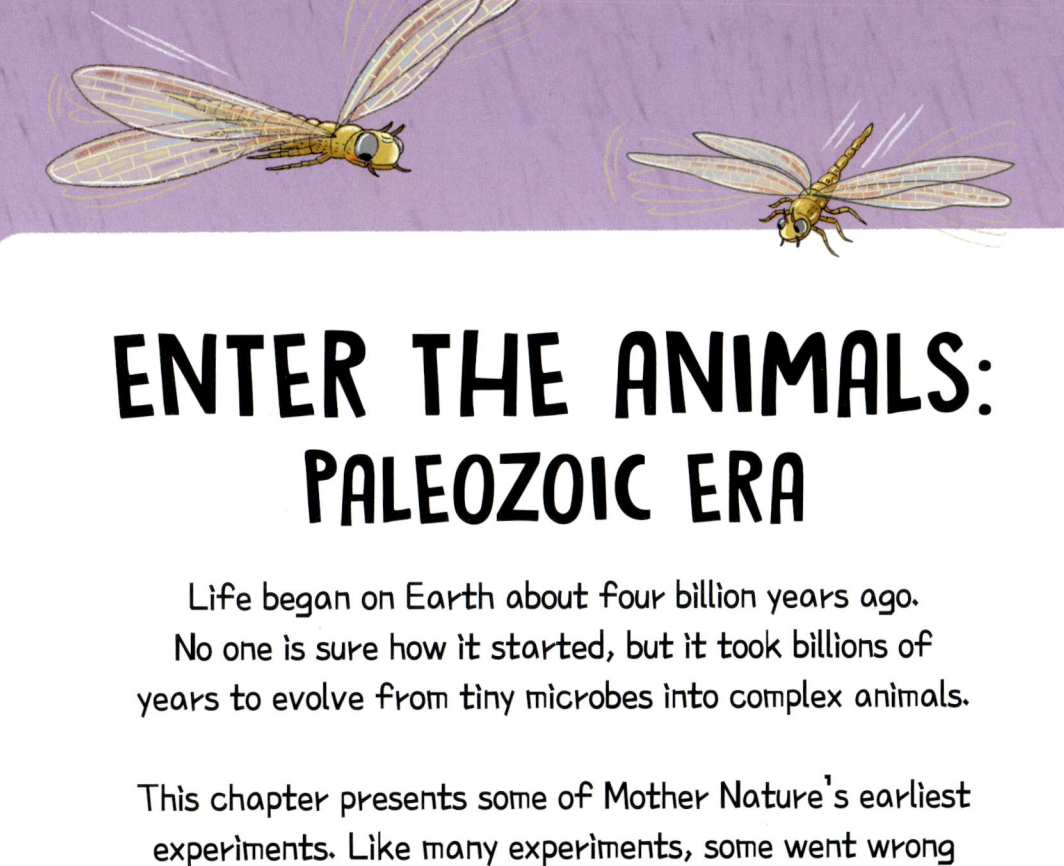

Life began on Earth about four billion years ago. No one is sure how it started, but it took billions of years to evolve from tiny microbes into complex animals.

This chapter presents some of Mother Nature's earliest experiments. Like many experiments, some went wrong and animals died out in a process known as 'extinction'.

Scientists are able to date fossils by looking at the rocks they are found in. Animals first appeared in the sea, in a distant time period called the Precambrian. A mere 200 million years later, Earth's early animals had taken their first steps on land.

Get ready to meet some of your very oldest relatives!

PERIODS AND ERAS

Periods are grouped together into longer stretches of time called 'eras'. Throughout this book, you can find out what period you're reading about by checking the top corner of the page. To see how all the different periods and eras flow together, take a look at the timeline on page 94.

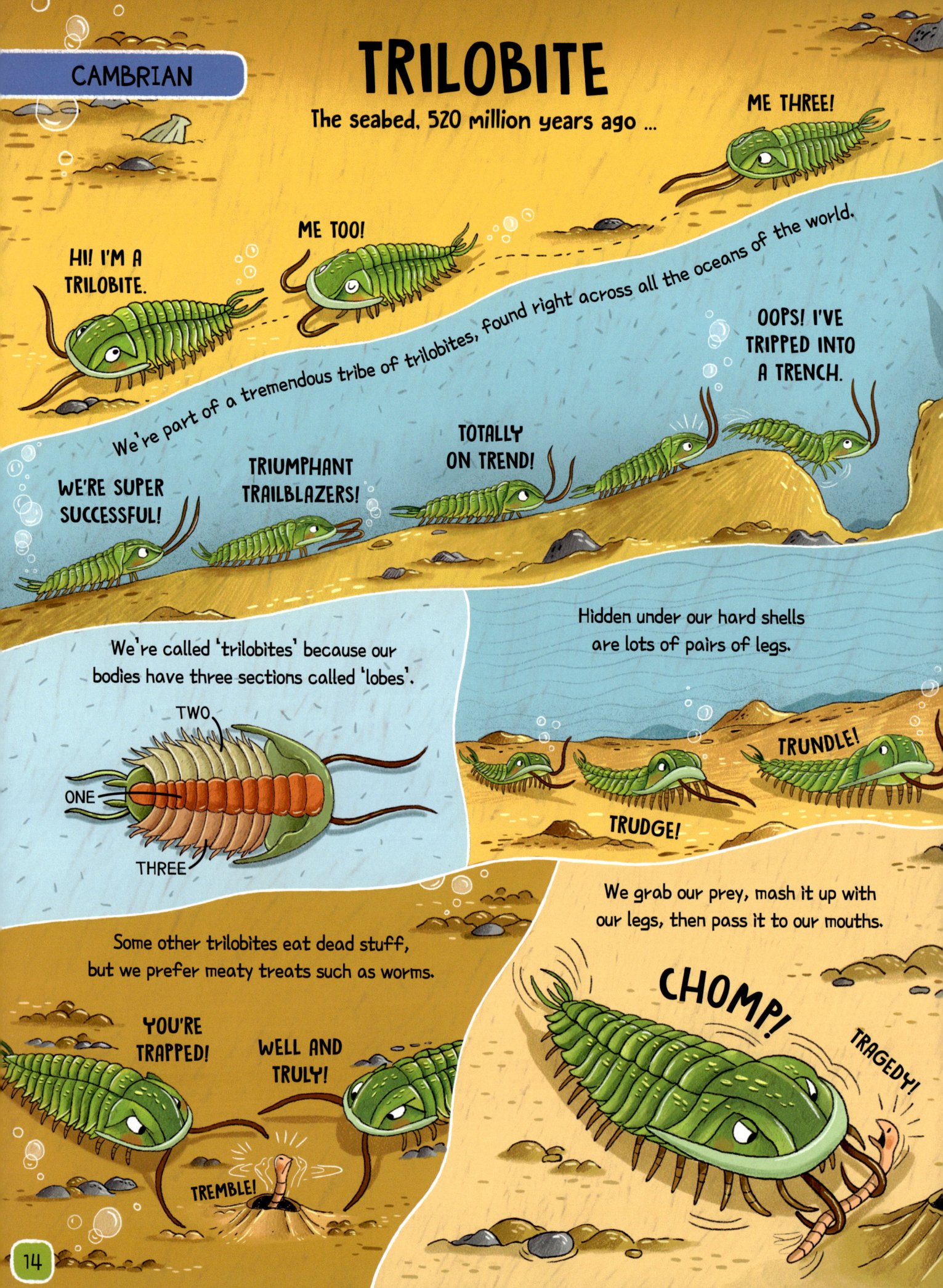

| CAMBRIAN | # OPABINIA | Beneath the surface of the sea, 510 million years ago... |

DEAD COOL	# CAMBRIAN CATWALK
	Life tried out lots of wild new looks during the Cambrian Period. Not all of them were successful, and some were super-strange!

HEADS OR TAILS?

Slug-like *Wiwaxia* was 5 centimetres long and covered in protective spines and scales. With no obvious head end, it was hard to tell if it was coming or going.

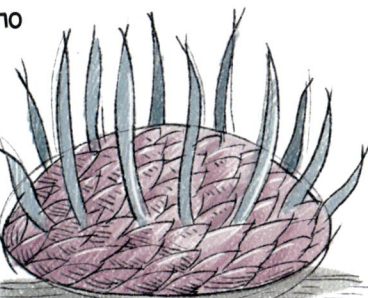

DREAM ON

Hallucigenia is well-named. Its bizarre body looked like something from a bad dream. Scientists originally thought the spikes on its back were legs!

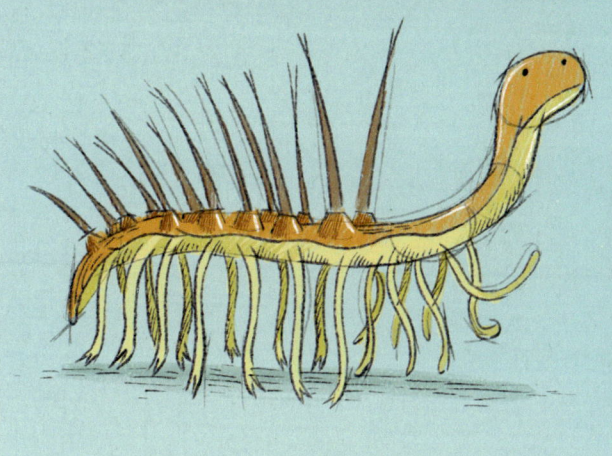

LEG UP

Marrella was one of the earliest arthropods*. It was about 2 centimetres long and used its legs not just to move, but also to breathe.

*Arthropods are animals with jointed legs, a segmented body and an exoskeleton (a hard covering, or 'outside skeleton'). Arthropods include crabs and insects.

GET A WIGGLE ON

Pikaia is a famous early animal. Its long, thin body had a form of primitive backbone that it may have used to swim like an eel.

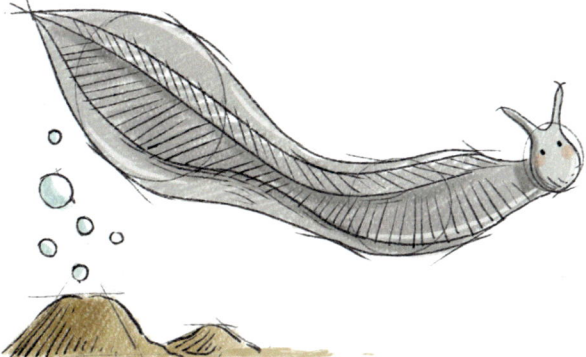

DEAD COOL	# DEEP DIVE
	Around 444 million years ago, the Ordovician Period ended — badly. Climate changes saw around half of all living things become extinct. Along the way, there were some animal fashions to die for.

BIG EATER

At over 2 metres long, *Aegirocassis* was the planet's biggest animal during the Ordovician. Despite its size, it fed on tiny floating plankton.

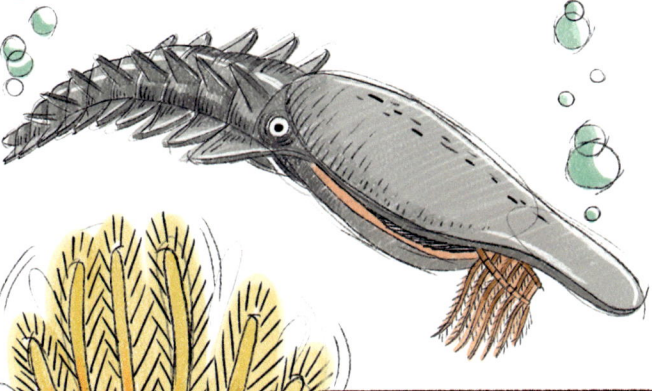

FIRST FISH

Arandaspis was the earliest known animal with a proper backbone. A primitive jawless fish, it had no fins and probably swam by wiggling its tail like a modern tadpole.

WIGGLE!

FLOWER POWER

Crinoids kicked off in the Ordovician Period. These filter-feeders with stalks, which are still around today, are related to starfish but look more like plants. That explains their nickname: 'sea lilies'.

HORN OF PLENTY

Endoceras was a top predator in the Ordovician Period, grabbing passing prey with its tentacles. This distant relative of the modern octopus had a cone-shaped shell that grew to over 2 metres in length.

COELACANTH

Deep in the ocean, over 390 million years ago ...

DEVONIAN

DEVONIAN

DUNKLEOSTEUS

Under the sea, some 380 million years ago ...

Hi! I'm a *Dunkleosteus*. Pleased to eat you ... er, I mean, 'meet' you.

I'll eat anything that crosses my path. Fish, trilobites, ammonoids (shells included). I love a lunch with a crunch!

FISH

TRILOBITE (SEE PAGES 14–15)

AMMONOID

At almost 9 metres long, I'm literally the big fish around here right now. See those shark-looking fish over there? I could eat them for breakfast.

NOT TODAY, YOU WON'T!

SO LONG, YOU BIG SUCKER!

CLADOSELACHE

Not only am I big, but my front end has thick, bony armour plating.

BONY HEAD

BUT DON'T YOU DARE CALL ME A 'BONEHEAD'!

Oddly, I don't have any teeth. Instead, I make do with these razor-sharp fangs made of bone.

WANT A CLOSER LOOK?

BONY FANGS

They sharpen themselves each time I open and close my mouth.

ARE YOU SURE YOU DON'T WANT A CLOSER LOOK?

DEVONIAN

TIKTAALIK

Close to the shoreline, 375 million years ago …

DEAD COOL	# GO FISH!
	'The Age of Fishes' was also an age of some funky new fashions. Here are some top Devonian designer looks.

FLAT TOP

Stethacanthus was a shark-like fish with an ironing board-shaped fin on its back, topped with spines. Only males had them, so maybe females found them fab!

NOSE JOB

Rolfosteus was a super-streamlined, armour-plated relative of *Dunkleosteus* (see pages 24–25). This 15-centimetre-long fish had large eyes and a long snout, which possibly helped it to sniff out its prey.

BREATH OF FRESH AIR

Dipterus was an early ancestor of a modern fish called a lungfish. It had gills, but could haul itself over land on its fins and breathe air. Look out, land — the lungs are coming!

FRONT ROW

With fingers rather than fins, *Ichthyostega* was a pioneering four-legged animal known as a 'fishapod'. It's thought it may have used its front legs to pull itself forward on land.

DEAD COOL

FASHIONABLY LATE

The Late Carboniferous Period finally drew to a close 300 million years ago. Here are four long-gone looks.

WATER WINGS

Squatinactis was an unusual shark-like fish that swam with 'wings', like a modern ray. It may have hidden on the seabed and leapt forward to seize passing prey.

MIGHTY BITER

At up to 3 metres long, *Ophiacodon* was a powerful land predator with a huge head. Its name means 'snake tooth', as it had jaws full of small, sharp teeth.

MESSY EATER

Eryops was a huge amphibian with heavy limbs, a bumpy body and a mouth full of backwards-facing teeth. These stopped prey from escaping its mouth and moved food down its throat.

LOOK, NO HANDS!

Some animals of the period lived on land but without any limbs, just like many of today's snakes. *Lethiscus* was a legless amphibian that is thought to have lived in burrows in the ground.

DIPLOCAULUS

A riverbed, 298 million years ago ...

PERMIAN

Just because I'm on a bed, doesn't mean I'm sleeping.

I'm a *Diplocaulus* — a very hungry amphibian from the Early Permian Period. Things here are finally looking up.

SWIM!

Come to think of it, I do a lot of looking up, given where my eyes are placed.

SWIM!

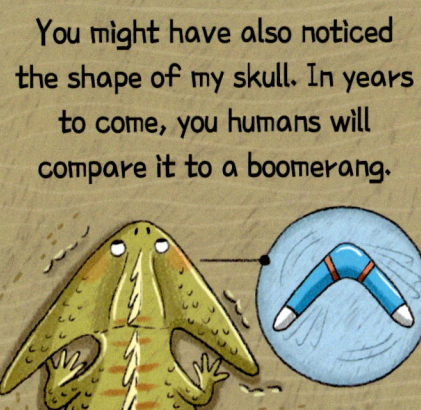

You might have also noticed the shape of my skull. In years to come, you humans will compare it to a boomerang.

SWIM!

ANY MOMENT NOW ...

That boomerang shape helps me to rise up quickly through the water and snap up my prey. Watch ...

LEAP! EEK!

And a wide head means a wide bite!

DOOMED!

Big Mouth strikes again!

After that I just drop down and return to where I was before.

FALL! BACK AGAIN!

Boomerang by name, boomerang by nature.

Ooh, here comes another one. Bye!

SWIM!

PERMIAN

DIMETRODON
A desert, 290 million years ago ...

"Hi! I'm a *Dimetrodon*. I'm 3 metres long and I'm the top predator right now!"

"Just look at all these different teeth of mine. I have long fangs for piercing my prey and smaller, saw-like ones for cutting it up. You really don't want to get on the wrong side of me!"

"In fact, you don't want to get on any side of me. Just look!"

- HUGE, FIN-LIKE BACK SAIL
- POWERFUL SKULL WITH STRONG JAW MUSCLES
- LONG TAIL FOR BALANCE
- SPRAWLING, LIZARD-LIKE STANCE
- 80 TERRIFYING TEETH

"My fin is made from extended bones in my spine with skin stretched over them."

"What's the fin for? Good question!"

DEAD COOL	# OVER AND OUT
	The Permian Period ended with most marine life and three-quarters of all land animals dying out, including these ill-fated four.

ARMOUR-GEDDON

Scutosaurus was a heavily armoured reptile with bony body plates and a spiky skull. Built like a slow-moving, plant-eating tank, its tough exterior protected it from predators.

THICK HEAD

Mighty *Moschops* was a vegetarian that possibly grazed in herds. It had skull bones 10 centimetres thick, and it's thought that rival *Moschops* battled by bashing their heads together.

FANG CLUB

Inostrancevia was a sabre-toothed predator with a seriously scary smile. Its terrifying top canine teeth grew up to 15 centimetres long.

LEAPING LIZARDS

Weigeltisaurus was a lizard-like animal capable of gliding flight thanks to an extendable membrane on either side of its body that acted like wings.

DINOSAUR DAYS: MESOZOIC ERA

The Permian mass extinction paved the way for the most famous and fearsome prehistoric animals of them all: dinosaurs.

Dinosaurs developed in a period known as the Late Triassic, and they and their reptile relatives dominated the land, sea and sky for over 150 million years.

The Mesozoic Era wasn't just about dinosaurs, either. It was also when the first birds took flight and some little furry things called mammals began to appear.

Dinosaurs grew bigger and beastlier throughout the Jurassic and Cretaceous Periods, but they were no match for the massive space rock that slammed into the Earth about 66 million years ago.

You know it ends badly, but as you'll see in this chapter, there were some amazing animals along the way.

TRIASSIC — # LYSTROSAURUS — A desert in Pangaea, 250 million years ago ...

Hi! I'm a *Lystrosaurus*. I'm a reptile about the size of a pig.

I'm one of the few species that survived the Permian extinction.

Luckily, these tasty seed-ferns survived, too. I love sinking my beak into them!

CHOMP!

I might be a relic from the Permian, but I never get lonely. Just look ...

I'm another *Lystrosaurus*!

Me too!

MUNCH!

And me!

CHOMP!

In fact, at least nine out of every ten land animals with a backbone alive right now is a species of *Lystrosaurus*. What are the chances?

WELL, THEY'RE AT LEAST 90 PER CENT.

I wasn't actually asking for an answer!

SIGH!

Because we're pretty big and there's so many of us, we don't really fear any predators. Not even this *Proterosuchus*.

WE SEE YOU!

BAH!

When I want to be alone, I can just pop back into my burrow.

HIDE!

It's hard to be alone when you're a *Lystrosaurus*.

HELLO AGAIN!

DEAD COOL

BOUNCING BACK

Life did eventually recover after the Permian disaster. The Triassic lasted 50 million years, with some powerful looks along the way. Here are four trademark Triassic tryouts.

NOODLE NECK

Tanystropheus was a fish-eating reptile with a really long neck. In fact, at up to 3 metres long, its neck made up over half of its body length. It may have had to live in water to support the weight of its head, or risk it snapping off!

TURNING TURTLE

Odontochelys was one of the world's first and most primitive turtles. It only had a hard shell on its lower side, possibly to protect it from predators lurking below.

BACK STORY

Longisquama was a 15-centimetre-long reptile with a fan of long scales on its back. It's thought the scales might have helped to give it some extra lift when it leapt into the air.

MAYBE MAMMAL

Morganucodon was a shrew-like animal from the late Triassic. Fossils suggest it had fur and fed milk to its babies (like a mammal), but laid eggs (like a reptile). That makes it similar to the modern duck-billed platypus.

LEEDSICHTHYS

Somewhere in the ocean, 170 million years ago ...

JURASSIC

* It's still unbeaten!

JURASSIC

ALLOSAURUS v STEGOSAURUS

A fern-covered plain, 153 million years ago ...

Hello! I'm an *Allosaurus*, a top predator. There are lots of us about.

EEK!

Hello! We're plant-eating *Stegosaurus*. There are lots of us, too.

CHOMP!

At up to 12 metres long and 5 metres tall, I'm one of the most massive meat-eaters around right now.

Being about the size of a modern elephant, we're among the most massive pieces of meat on offer.

THAT SOUNDS RISKY!

My mighty hind legs help me hunt huge herbivores, which I can then grip with my killer front claws.

I'M OFF!

15 CENTIMETRES LONG

Our long back legs tip our bodies down towards the ground.

ALL THE BETTER FOR BROWSING WITH OUR BEAKS!

CHEW!

My giant head is packed with 32 teeth shaped like steak knives. And when they fall out, they just grow back even sharper!

PING!

Our heads are tiny compared to our big, bulky bodies.

AND OUR BRAINS ARE EVEN SMALLER!

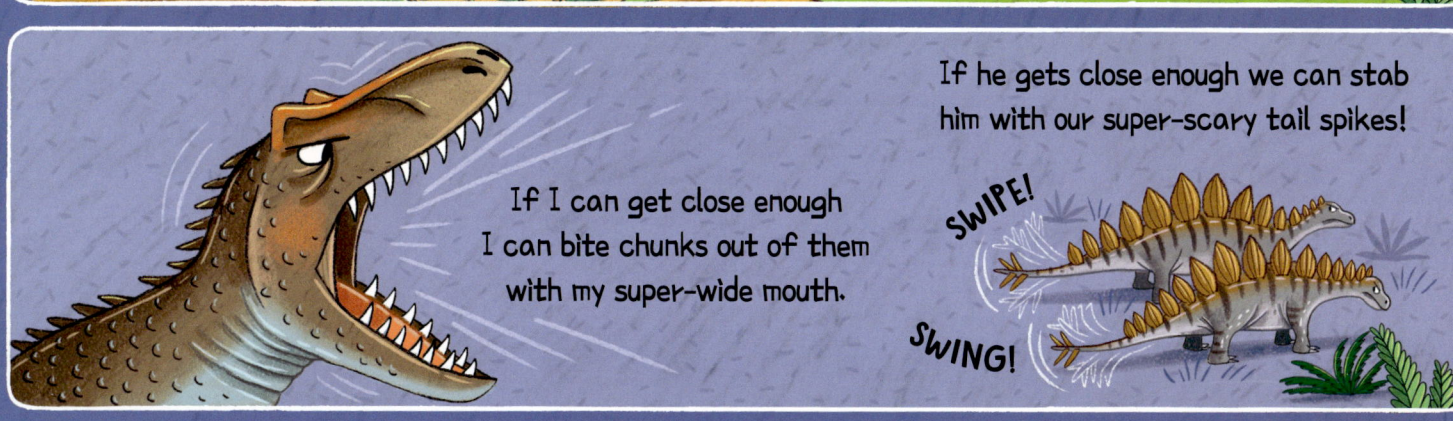

BRACHIOSAURUS

A flat plain, 152 million years ago ...

JURASSIC

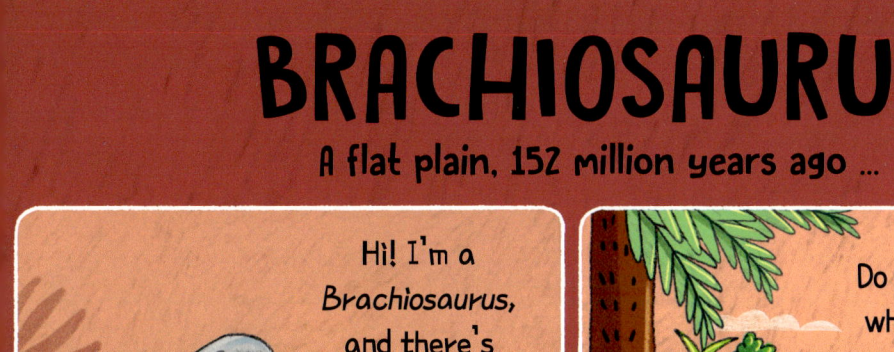

Hi! I'm a *Brachiosaurus*, and there's something very unusual about me.

Do you know what it is?

IS IT YOUR FRONT LEGS?

JUVENILE

Yes, it is! Like a modern-day giraffe, my front legs are longer than my back ones.

They tilt me up to the tallest treetops. Luckily, I have a head for heights.

With my 9-metre-long neck, I tower above other plant-eaters like this *Apatosaurus*.

They have to rear up to match me. If I tried that, I'd fall over!

WOBBLE!

I can also clear all the plants around me when I bend down.

GONE!

All I have to do is rake them in with my teeth.

TUG!

Of course, there are drawbacks to having legs like these. Hills are too hard for me to climb.

THINK I'LL GIVE THIS A MISS.

So I stick to wandering across flat plains in a big family group, looking for new plants to eat.

ARE WE NEARLY THERE YET?

MUM, I NEED A WEE!

TRUDGE! TREK! TRUDGE!

49

JURASSIC

DIPLODOCUS

A forest's edge, 152 million years ago...

Hello! I'm a *Diplodocus*. There's more to me than meets the eye.

That said, at 33 metres long, there's a lot of me that does meet the eye, too!

SWISH!

Sure, I'm big, but I'm not the only one. Check out these giants who are also around at the moment.

GIRAFFATITAN — 26 METRES LONG
TURIASAURUS — 30 METRES LONG
SUPERSAURUS — 34 METRES LONG
MAMENCHISAURUS — 35 METRES LONG

HIYA!

To maintain a body like mine, I need to eat a lot of greens.

You might even say my diet is magic. Watch this!

BITE! RAKE! GONE!

These peg-like front teeth of mine make foliage simply disappear!

I don't just eat greens. I also eat browns and greys. Look!

GRAVEL

See? Those stones have disappeared, too.

GONE!

RHAMPHORHYNCHUS

JURASSIC

A rock by the sea, 147 million years ago ...

Hi! I'm a flying reptile called *Rhamphorhynchus*. LET'S GO FISH!

My wings are a membrane stretched between my legs and my extra-long fourth fingers. LEAP! WE HAVE LIFT-OFF!

See my teeth? They're great for snatching fish from the surface of the water.

I spy a shoal. I'm going in! SWOOP!

Look at this little *Leptolepides* fish just swimming along near the surface.

Too near the surface, it seems! GRAB! EEK! QUICK! DIVE!

Uh-oh, that *Aspidorhynchus* looks hungry. And big! LUNGE!

BITE! Help! The fish are fighting back!

Get off me! I CAN'T! I'M STUCK IN YOUR WING!

I've got a sinking feeling. SERVES YOU RIGHT!

A FAMOUS FOSSIL FOUND 150 MILLION YEARS LATER ... I'm still cross about this. ME TOO.

DEAD COOL	# JURASSIC SPARKS
	Nature got back to its inventive best in the Jurassic Period, and it wasn't just about huge dinosaurs. There were plenty of smaller animals going around, too, such as these four.

SCRATCH THAT

Warning: reading this may make you itch! *Pseudopulex* was a primitive flea, 50 times bigger than a modern dog flea. It had a mouth like a hypodermic needle, and fed on the blood of dinosaurs!

HOP TO IT

Prosalirus was one of the first known frogs. Frogs swapped the long tails of earlier amphibians for strong back legs that — literally — allowed them to make a great leap forward!

WHAT THE DEVIL?

Gryphaea were super-successful, sea-dwelling molluscs that lived inside curved shells, like curly oysters. Commonly found as fossils, their shape gave rise to their nickname: 'devil's toenails'!

DIG THIS

Fruitafossor was a small mammal, the size of a modern hamster. It's thought to have munched on termites, and had muscly forearms that it probably used for digging.

CRETACEOUS

AMARGASAURUS

A forest in what is now South America, 125 million years ago ...

Hello! We're a herd of hungry herbivores called *Amargasaurus*.
MUNCH!
CRUNCH!

We're keeping our heads down, in case any large predators spot us.
MUNCH!

We can't really raise our heads very high anyway. These spines get in the way.
ANY LUCK?
I'M TRYING!
STRAIN!

Luckily, our food is mostly low down. Like these scrummy cycads ...
YUM!

These two rows of spines on our necks may look odd, but they have their uses.

Take these hungry little meat-eating *Ligabueino*, for example.
WE SEE YOU!
OOH, LUNCH!

Everybody, bend!
WHAT?

CURL!

Attack us, and you'll be the ones getting a pain in the neck.
BAH!

Now, back to that food.
FRESH FERNS! THINGS ARE LOOKING UP!
UNLIKE US!

LATER ... It's getting dark. Keep your heads down!

WE CAN'T DO ANYTHING ELSE!

SPINOSAURUS

A swamp in what is now Africa, 95 million years ago ...

CRETACEOUS

57

CRETACEOUS

PTERANODON

A rocky island in a shallow sea, 85 million years ago...

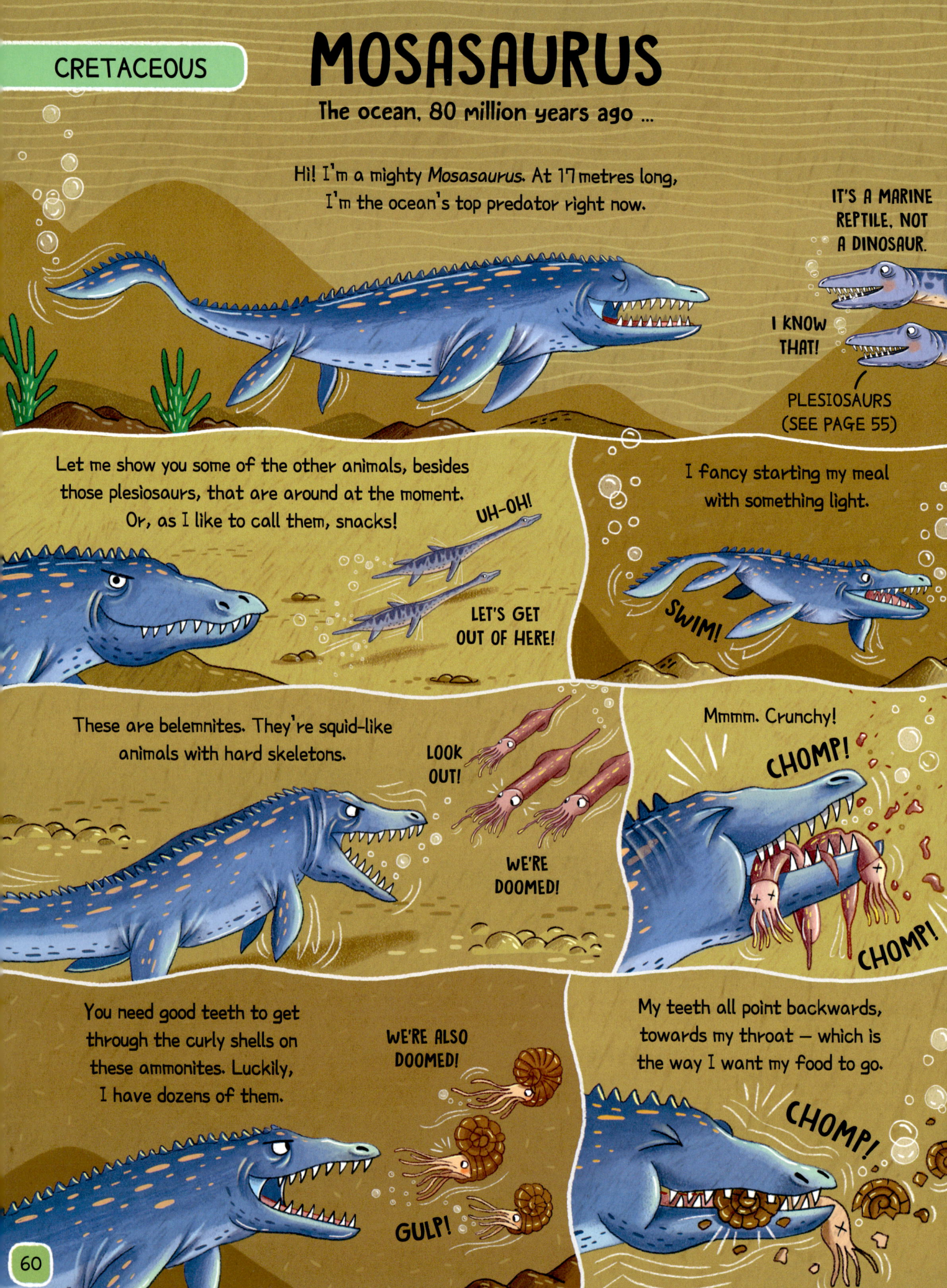

VELOCIRAPTOR

A dry, dusty desert in what is now Mongolia, 73 million years ago ...

CRETACEOUS

CARNOTAURUS

A forest in what we now call Argentina, 70 million years ago ...

CRETACEOUS

Hi! I'm a *Carnotaurus*. The name means 'meat-eating bull'. And with these teeth, it's pretty obvious that I eat meat.

The 'bull' bit comes from these distinctive horns, of course. Why do we have them? Well ...

Maybe us males use them for fighting over females.

MY HORNS ARE BIGGEST!

NO, MINE ARE!

And just look at these legs. Impressive, right?

Just don't look at my tiny arms.

OI! I SAID DON'T LOOK AT THEM!

KRARK! SOMETIMES I WISH I COULD ACTUALLY ROAR.

Anyway, right now I'm lying in wait. I'm going to ambush some unsuspecting plant-eater and sink my jaws into its side.

OOH! SOUNDS LIKE I'M IN LUCK.

STOMP! STOMP!

But ... um ... some plant-eaters are easier to ambush than others.

YOU WEREN'T WAITING FOR US, WERE YOU?

DREADNOUGHTUS (AT 26 METRES, ONE OF THE LONGEST DINOSAURS EVER)

Time to stretch these legs, I suppose.

HE HAD HORNS. HOW UNUSUAL!

I ONLY NOTICED HIS TINY ARMS.

ANKYLOSAURUS

CRETACEOUS

A woodland in what is now North America, 68 million years ago …

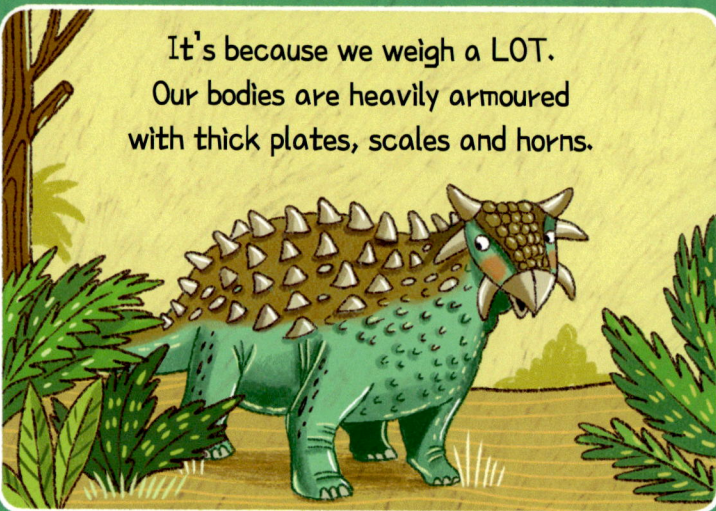

CRETACEOUS

TYRANNOSAURUS v TRICERATOPS
A forest in what is now North America, 66 million years ago ...

Hi! Me again. I'm the *Tyrannosaurus rex* from pages 66 and 67. My name means 'king tyrant lizard'. Although I'm not a lizard, it pretty much sums me up. I rule here!

Hi! I'm a *Triceratops*. My name means 'three-horned face'. It's not exactly flattering, but it is accurate.

Us *T. rex* females are bigger than the males. At 12 metres long, I'm one of the most terrifying land animals ever, with a horrifying hiss to match.

HISSS!

That *T. rex* likes to boast, but I'm the real big head. I'm a male, and my skull makes up a third of my body. Ladies seem to like it, though.

LOOKING GOOD!

Marvel at my mighty mouth. I have 60 teeth the size of bananas (whatever they are). The bones my teeth crunch through end up in my poo.

WHIFF!

Pff! So what? With my horny beak, I can wreak havoc on these ferns. Plus, my poo helps spread their seeds. It's poo-tiful!

CHOMP! WHIFF!

My poo isn't the only thing that stinks. All this rotten meat between my teeth makes my breath as bad as my bite!

PONG! STENCH!

Yuck! I can smell rex-breath. There must be one nearby. I'm off!

SNIFF! SNIFF!

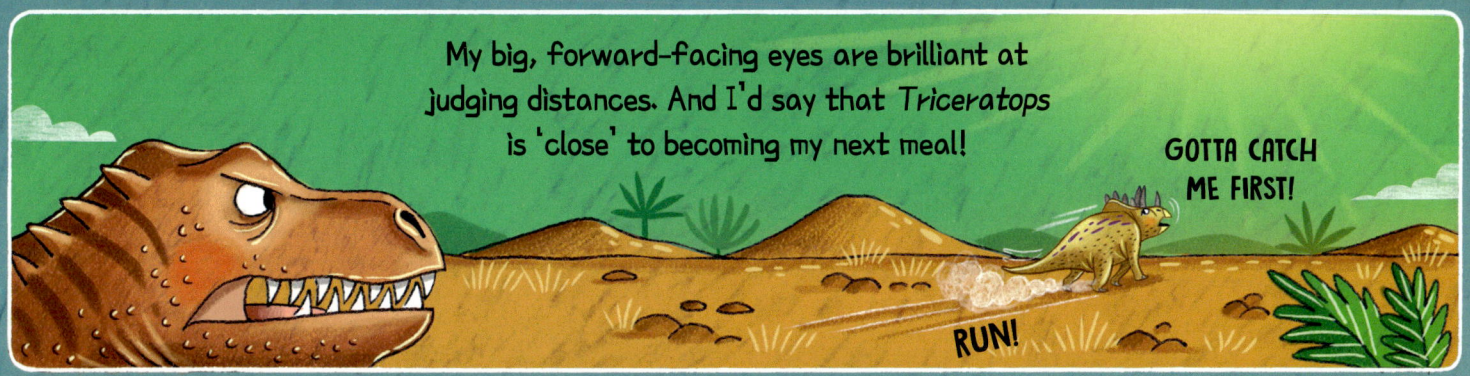

DEAD COOL

AUDACIOUS CRETACEOUS

The Cretaceous Period ended with a bang when an asteroid the size of Paris smashed into the planet, close to what is now known as the Gulf of Mexico. Here's a last look at some Cretaceous crew.

OUT OF AMMO

All the ammonites died out, including *Diplomoceras*, with its paperclip-shaped shell, and the curly-shelled *Didymoceras*, which looked like some sort of strange musical instrument.

BIG MOUTH

Nigersaurus was a dinosaur lost from what is now Africa. It had a super-wide smile, with 500 teeth in rows at the front for grazing on greens. Imagine having to brush all those!

KILLER NEWT

Koolasuchus was a car-sized amphibian from the early Cretaceous. It lived in what is now Australia and dined on dinosaurs, catching them with a fang-packed head the size and shape of a dustbin lid.

GO APE!

Purgatorius was a squirrel-like mammal that survived the Cretaceous catastrophe. Its teeth suggest it may have been an early ancestor of apes — and us!

AFTER THE ASTEROID: CENOZOIC ERA

With the dinosaurs wiped out, birds — their close relatives — were left to carry on their legacy. However, it was those small, shrew-like mammals that were to make it big. Massive, in fact.

Huge, hairy beasts feasted on flourishing grasses and flowering plants. Those herbivores were hunted down by powerful predators with bone-crushing jaws and giant, curving canine teeth.

While some mammals were 'mammoth' by both name and nature, it was smaller, ape-like animals that were to have the most lasting impact — as the doomed dodo would learn to its dismay.

This chapter brings the animal kingdom bang up to date. Who do you think should rule it now, and can we learn from our past? Read on to decide.

PALEOGENE

AMBULOCETUS

An ocean edge in what is now India, 48 million years ago ...

I'm a 3-metre-long mammal called *Ambulocetus*. The name means 'walking whale' ...

HEAVE!

... though I'm not sure you could really call this 'walking'.

HEAVE!

Frankly, I find moving on land to be a real drag. Literally!

DRAG! HEAVE!

However, once I'm in water, I'm a beautiful sight ... though this tasty-looking fish might not agree!

SWIM! EEK!

I swim by pulsing my body up and down.

PULSE! STOP FOLLOWING ME!

Scientists in your time say my jaws show that I'm an ancestor of modern whales.

Just like modern whales, I use the bones in my jaws to help me hear underwater.

CAN'T YOU JUST USE THEM FOR THAT, THEN?

SNAP!

My jaws are also full of teeth for catching fish, as you can see.

ALL TOO CLEARLY!

Sorry, I can't hear you over the noise of me eating.

CHOMP! SWALLOW!

I'm having a whale of a time!

BURP!

PALEOGENE — ARCHAEOTHERIUM

A grassy, open forest in what is now North America, 33 million years ago ...

Hi! I'm an *Archaeotherium*. With these tusks, you might think I look like a big, scary, killer pig!

But am I a pig? Good question. Right now, I can see a herd of grazing *Poebrotherium*. They're not pigs. In fact, they're closer to modern camels.

MUNCH!
CHOMP!

And that thing I can see creeping up on them is a meat-eating mammal called a *Hyaenodon*. It's not a pig either.

STALK!

Those *Mesohippus* over there have spotted it, too. They're not pigs. They're an early form of horse.

WE'RE OFF!
GALLOP!

This *Hyracodon* is also running away. It's only 60 centimetres tall. And it's also not a pig. It's a type of primitive, zippy hippo.

RUN!

I hope you're keeping up with this. Just like the *Hyaenodon* is trying to keep up with that *Poebrotherium* herd now that they've spotted him.

THESE EARLY CAMELS ARE GIVING ME THE HUMP!

LIMP!

What that injured, limping *Poebrotherium* at the back doesn't know is that a second *Hyaenodon* is lying in wait ...

GRRR!
HELP!
LIMP!
RUN!
HA!

NEOGENE

MEGALODON

Any of the world's warm oceans, about 20 million years ago ...

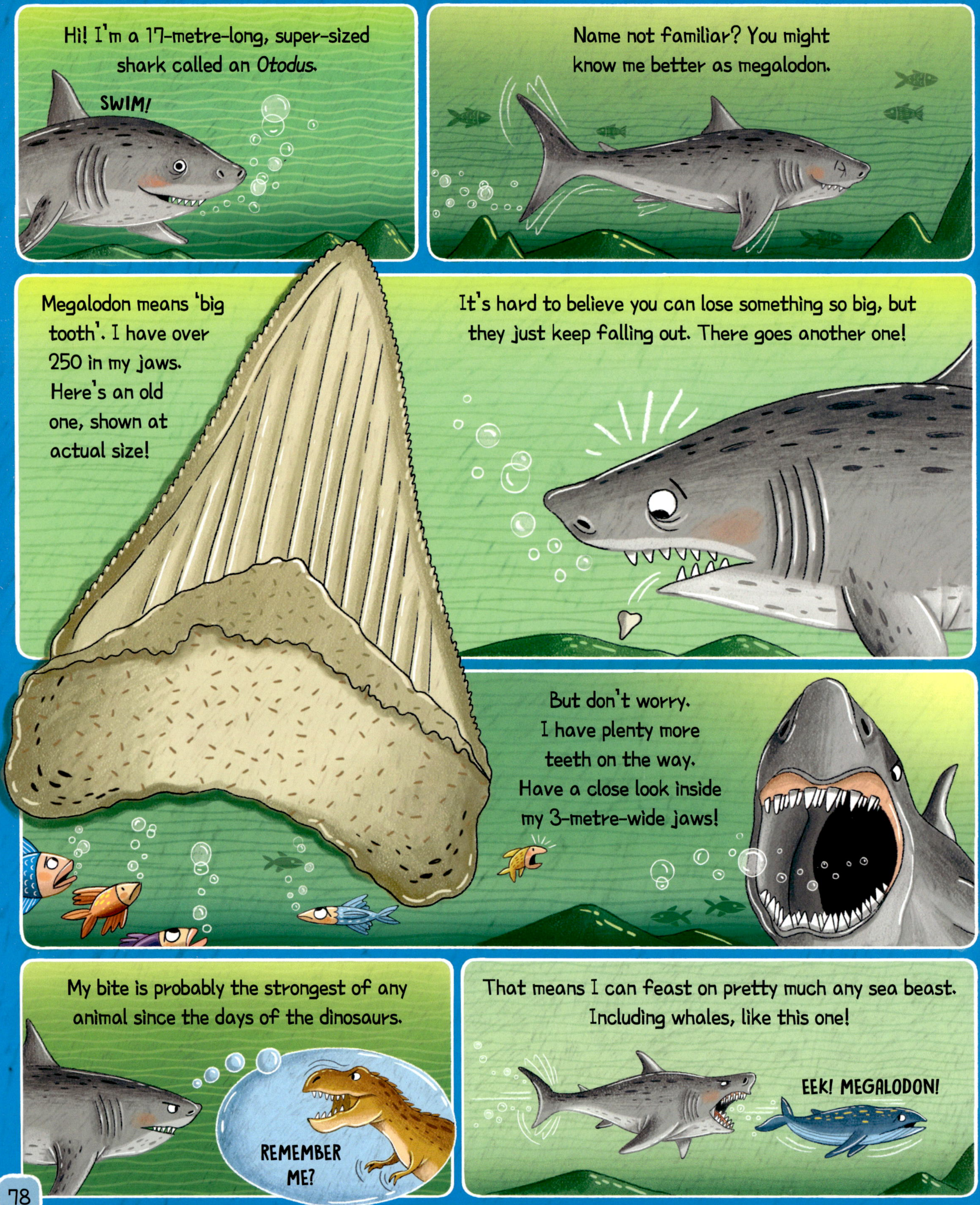

Hi! I'm a 17-metre-long, super-sized shark called an *Otodus*. SWIM!

Name not familiar? You might know me better as megalodon.

Megalodon means 'big tooth'. I have over 250 in my jaws. Here's an old one, shown at actual size!

It's hard to believe you can lose something so big, but they just keep falling out. There goes another one!

But don't worry. I have plenty more teeth on the way. Have a close look inside my 3-metre-wide jaws!

My bite is probably the strongest of any animal since the days of the dinosaurs. REMEMBER ME?

That means I can feast on pretty much any sea beast. Including whales, like this one! EEK! MEGALODON!

NEOGENE — PHORUSRHACOS

A grassland in what is now Argentina, 15 million years ago …

Hi! I'm a mole-like mammal called a *Necrolestes*.

I'm in the wrong place. This page isn't about me.

So, I'm off. Bye!
BURROW!

MEANWHILE, NEARBY …
Shhh! I'm a 2.5-metre-tall 'terror bird' called a *Phorusrhacos*.

I'm keeping my voice down and pretending to be a tree, so this rabbit-like snack called a *Pachyrukhos* comes close enough for me to grab it.
NIBBLE!

That's what my beak and killer claws are for.
WHAT?

Oops! Spoke too loudly. Those ears hear everything! I might not be able to fly, but with these legs I can certainly run!
HOP! HOP! ME TOO!

Now it's time to put that beak into action.
LUNGE!
SNAP! SNAP!
EEK!

Bah! Missed!
HIDE!

Oh well. Back to being a tree.

LATER … Remember me?
DIG! BURROW!

Time to surface and see where I am. Going up!

Oh no … wrong place AGAIN!

DEAD COOL	# THE FUR BRIGADE
	Mammals made the most of the post-dinosaur world. Some grew huge. Some grew horns. Some stayed small, but grew horns anyway! Here are four fab mammals of the Cenozoic Era.

TUSK FORCE

Deinotherium was shaking the ground of Africa and Europe 10 million years ago. It was twice as heavy as a modern African elephant and had tusks on its chin, possibly for removing branches so it could get to more tasty leaves.

MINI MONKEY

Eosimias was an early ancestor of modern monkeys (and also humans) that lived in China over 40 million years ago. It was seriously small, with a body about 12 centimetres long.

GOPHER IT

At 30 centimetres long, gopher-like *Ceratogaulus* is the smallest horned mammal ever discovered. It lived in burrows some 10 million years ago, and most likely used its horns for holding off prying predators.

MEAL DEAL

Amphicyon was a massive meat-eater found widely across the globe. Looking like a mix of a bear and a dog, it had terrifying teeth, but was probably an omniovore, meaning it ate plants as well as prey.

AUSTRALOPITHECUS

Southern Africa, 3 million years ago ...

NEOGENE

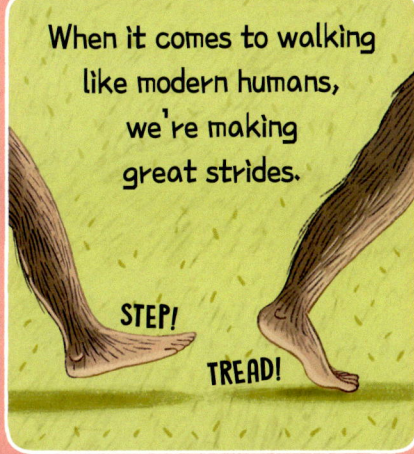

QUATERNARY

MEGATHERIUM

A grassland in what is now Bolivia, South America, 2 million years ago...

Hi! I'm a giant, grazing ground sloth called a *Megatherium*.

MUNCH! CHOMP!

Sloth means 'moving slowly' and *Megatherium* means 'great beast', so my name's pretty accurate. I'm over 6 metres long!

TRUDGE! MUNCH!

I'm not just long. I'm tall, too! I can stand up like this, you see. I use my tail for balance, like a living tripod.

In fact, I may be the largest mammal ever to have walked on two legs.

CHOMP! TRUNDLE!

That means I can reach tasty, high-up leaves like these ones. I use my huge claws to bend down the branches.

What's this? Ah, a *Glyptodon* — a massive armadillo. Because of its big, heavy shell, it's only able to graze down here at ground level.

YEAH, BUT HEAVY SHELLS HAVE THEIR USES.

QUATERNARY

WOOLLY MAMMOTH V NEANDERTHAL

Southern Europe during the last ice age, about 40,000 years ago...

We're woolly mammoths. As our name suggests, we're furry and we're massive!

OUR FUR IS OVER 30 CENTIMETRES LONG IN SOME PLACES.

We're a species of human called Neanderthals. We wear these furry animal skins to keep warm.

CAN YOU GUESS WHICH ANIMAL WE GOT THEM FROM?

We're similar in size to a modern African elephant, but with smaller ears to stop them from freezing, and thick fat for insulation.

A THICK, FURRY BUM FLAP HELPS, TOO!

Our big, broad noses help to warm the cold air before it enters our lungs. Oh, and we have fire, too!

WHOEVER INVENTED THIS WAS A GENIUS!

CRACKLE!

Our tusks grow up to 5 metres long. They're helpful for clearing the snow to get to the plants underneath.

SWIPE! CLEAR!

We find mammoth tusks useful, too. We use tusks and mammoth skins to build shelters for our hunting expeditions.

I MISS THE FIRE.

ME TOO.

We travel in herds to look for new places to graze, with our big female leading the way.

CAN WE SLOW DOWN? I'M ONLY LITTLE!

NO YOU'RE NOT - YOU'RE MAMMOTH!

We've learned to lie in wait, with our flint-tipped spears ready to ambush the herd. Doing all this shows we have considerable intelligence, and the advanced use of language.

WELL, WE DO HAVE BIGGER BRAINS THAN MODERN HUMANS.

UG!

QUATERNARY

DIRE WOLF

The Rancho La Brea tar pits in California, USA, 15,000 years ago ...

Hi! We're a pair of dire wolves.
AND I'M A GREY WOLF.

Us dire wolves are bigger and stronger than grey wolves.
GRRR!
GULP!

And we have the strongest bite of any member of the dog family, ever!
SNARL!
I'LL BE OFF, THEN!

Dire means 'terrible' ... and we're both terribly hungry right now.
DROOL! DROOL!

We're terribly lucky, too! We've found the corpse of a bison that must have got stuck in a pool of water.

So now, we feast!
LEAP!

Uh-oh. This pool isn't full of water after all. It's tar!
STUCK!

Well, this is awkward.
ALSO STUCK!

LATER ...
WHAT ARE OUR CHANCES OF GETTING OUT OF THIS?
DIRE.

EVEN LATER ...
WHAT A LUCKY LION I AM. TWO DEAD WOLVES AND A BISON TO FEAST ON.

WELL, THIS IS AWKWARD.
STUCK!

FIRST EXCAVATED OVER 100 YEARS AGO, THE RANCHO LA BREA TAR PITS CONTAINED REMAINS OF ABOUT 200,000 ANIMALS, INCLUDING HUNDREDS OF DIRE WOLF SKULLS.

DODO

The remote island of Mauritius in the Indian Ocean, in the year 1600 CE ...

QUATERNARY

DEAD COOL	# LIFE LESSONS
	Extinction is natural. Animals have been coming and going for billions of years, though today humans have a hand in it, too. These four were lost in the last century or so ...

END OF THE LINE

The passenger pigeon was once the most abundant bird in North America. Hunting saw its numbers fall from three billion to a single female, Martha, who died in a zoo in 1914.

TIGER FEAT

The thylacine, also called the Tasmanian tiger, was a meat-eating marsupial native to Australia. The last-known thylacine died in 1936, but some people insist they have seen them since. Has the Tasmanian tiger survived?

END OF THE TOAD

The golden toad lived in cloud forests high up in Costa Rica. They were tiny (only around 40 millimetres long) but beautiful. The last golden toad was spotted in 1989. Have they all croaked?

BACK TO LIFE?

The Pyrenean ibex, a large, horned European goat, went extinct as recently as the year 2000. Why it died out is uncertain, but scientists have been trying to bring it back, using cells taken from the last living animal — a process known as 'de-extinction'.

LIVING LEGENDS

HUMAN

HELLO! My scientific name is **HOMO SAPIENS**, but I'm better known as a **HUMAN**.

If you think us humans are the planet's most successful animal, **THINK AGAIN**. We've only been around for roughly **200,000 YEARS**. By comparison, **T. REX** managed about two million years and **JELLYFISH** go back around 600 million years. All of these living things are **CONNECTED**, and modern humans carry clues that lead back to our ancient ancestors. Here are some you can **SPOT FOR YOURSELF!**

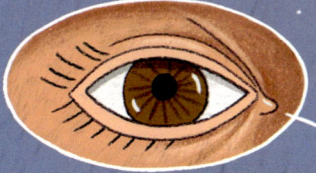

That **LITTLE FOLD** in the corner of each eye is all that remains of a former **THIRD EYELID**, still found in birds, reptiles and fish.

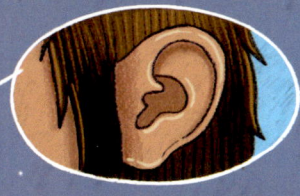

Although some of us can **WAGGLE OUR EARS**, humans are no longer able to angle their ears towards sounds. However, the **MUSCLES** for doing this are still there.

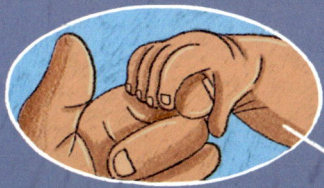

Newborn babies will instinctively **GRIP AN ADULT'S FINGER**. This reflex is also found in baby monkeys, who **GRIP THEIR MUMS' FUR** for safety.

Getting **GOOSEBUMPS** and **RAISED HAIRS** when scared is an **ANCIENT REFLEX** that would have made the bodies of our much **HAIRIER ANCESTORS** look larger to predators.

All humans develop a **TINY TAIL** for a few weeks before they are born. The bones of this tail, called the 'coccyx', are **STILL THERE** at the end of your spine.

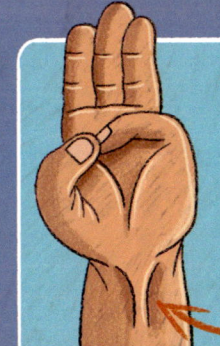

Touch your little finger to your thumb. Six out of seven people will see a **ROPE-LIKE TENDON** raising up on their wrist. That tendon is the remains of a muscle that helped our ancestors to move through trees. If you don't have it, **DON'T WORRY** — you don't need it!

GLOSSARY

Here are explanations of some words you may have read while meeting the animals in this book.

AMPHIBIAN
A scaleless, cold-blooded vertebrate animal living in water and on land. Many amphibians, such as frogs and toads, start out as tadpoles.

BIRD
A warm-blooded vertebrate animal with feathers. Now regarded by most scientists as technically dinosaurs themselves.

CARNIVORE
Any animal (or plant) that eats meat.

COLD-BLOODED
An animal whose body temperature depends on the temperature of its surroundings. It is sluggish when it is cold, and active when its surroundings are warm.

CYCAD
A woody, palm-like type of plant.

DE-EXTINCTION
The scientific process of attempting to bring back animal species that have died out.

DINOSAUR
A group of animals that went extinct around 66 million years ago. Modern birds are their living descendants.

EXTINCTION
The death of all living members of a plant or animal species, resulting in that species' complete disappearance from Earth.

FISH
A vertebrate animal, usually cold-blooded, that has a tail, fins and gills and can live and breathe underwater.

FOSSIL
Preserved traces or remains of living things from many years ago. Scientists can use fossils to learn about different species that died out before humans were around.

HERBIVORE
An animal that eats only plants.

INVERTEBRATE
An animal without a backbone.

MAMMAL
A vertebrate animal that feeds its young with milk. Most (but not all) mammals are born as babies rather than hatching from eggs, and have hair or fur. Humans are mammals.

OMNIVORE
An animal that eats both meat and plants as part of its diet.

PANGAEA
A huge 'supercontinent' that existed on Earth during the late Paleozoic and early Mesozoic Eras. Pangaea began to break apart around 200 million years ago, eventually forming the continents and oceans we have today.

PREDATOR
A living thing that catches animals in order to eat them.

PREHISTORIC
Anything dating back to a time before humans developed the ability to write and make historical records.

REPTILE
A cold-blooded animal that has scaled skin and lays eggs.

VERTEBRATE
An animal with a backbone.

WARM-BLOODED
An animal that can regulate and maintain its own temperature, regardless of its surroundings.

TIMELINE

Planet Earth is pretty old. Over 4.5 billion years old, in fact. The Earth's life story (and this book) is divided up into periods, from the watery world of the Precambrian all the way through to today. Here are the ages of Earth.

CENOZOIC ERA Pages 71–91	**QUATERNARY** 2.6 million years ago to today
	NEOGENE 23 to 2.6 million years ago
	PALEOGENE 66 to 23 million years ago
MESOZOIC ERA Pages 39–70	**CRETACEOUS** 145 to 66 million years ago
	JURASSIC 201 to 145 million years ago
	TRIASSIC 252 to 201 million years ago
PALEOZOIC ERA Pages 9–38	**PERMIAN** 299 to 252 million years ago
	CARBONIFEROUS 359 to 299 million years ago
	DEVONIAN 419 to 359 million years ago
	SILURIAN 444 to 419 million years ago
	ORDOVICIAN 485 to 444 million years ago
	CAMBRIAN 541 to 485 million years ago
	PRECAMBRIAN 4.6 billion to 541 million years ago